Editor
Lorin Klistoff, M.A.

Editorial Manager
Karen J. Goldfluss, M.S. Ed.

Editor in Chief
Sharon Coan, M.S. Ed.

Cover Artist
Jessica Orlando

Art Coordinator
Denice Adorno

Creative Director
Elayne Roberts

Imaging
Stephanie A. Salcido
James Edward Grace

Product Manager
Phil Garcia

Publishers
Rachelle Cracchiolo, M.S. Ed.
Mary Dupuy Smith, M.S. Ed.

How t Word Prodiems

Grades 1–2

Author

Mary Bolte

Teacher Created Materials, Inc.
6421 Industry Way
Westminster, CA 92683
www.teachercreated.com

ISBN-1-57690-941-7

©2000 Teacher Created Materials, Inc.
Reprinted, 2000
Made in U.S.A.

Table of Contents

A Note to Teachers and Parents

Welcome to the "How to" math series! You have chosen one of over two dozen books designed to give your children the information and practice they need to acquire important concepts in specific areas of math. The goal of the "How to" math books is to give children an extra boost as they work toward mastery of the math skills established by the National Council of Teachers of Mathematics (NCTM) and outlined in grade-level scope and sequence guidelines.

The design of this book is intended to be used by teachers or parents for a variety of purposes and needs. Each of the units contains one or more "How to" pages and two or more practice pages. The "How to" section of each unit precedes the practice pages and provides needed information such as a concept or math rule review, important terms and formulas to remember, or step-by-step guidelines necessary for using the practice pages. While most "How to" pages are written for direct use by the children, in some lower-grade level books, these pages are presented as instructional pages or direct lessons to be used by a teacher or parent prior to introducing the practice pages.

About This Book

How to Solve Word Problems: Grades 1–2 presents a comprehensive overview of word problems for students at this level. It can be used to introduce and teach basic word problems to children with little or no background in the concepts. This book can also be used in a learning center containing materials needed for each unit of instruction.

The units in this book can be used in whole-class instruction with the teacher or by a parent assisting his or her child through the book. This book also lends itself to use by a small group doing remedial or review work on word problems or for individuals and small groups engaged in enrichment or accelerated work. A teacher may want to have two tracks within his or her class with one moving at a faster pace and the other at a gradual pace appropriate to the ability or background of the children.

The word problem activities in this book will help your children learn new skills and reinforce skills already learned, as they learn how to:

- use mathematical content to explore and understand the solutions to word problems.
- create word problems from everyday situations.
- become confident in using math in a meaningful way.
- relate everyday language, pictures, and diagrams to mathematical language, symbols, and ideas.
- reflect upon and explain thinking about mathematical ideas and situations.
- use patterns to analyze mathematical situations.
- link conceptual and procedural knowledge.
- explore and apply estimation strategies.
- understand the numeration system through counting, grouping, and place value concepts.
- develop operational sense.
- develop proficiency in basic fact computation and techniques.
- recognize and appreciate geometry in the world.
- use time measurement in everyday situations.
- apply fractions to problem situations.

If children have difficulty with a specific concept or unit within this book, review the material and allow them to redo the troublesome pages. It is preferable for children to find the work easy at first and to gradually advance to the more difficult concepts.

How to Solve Word Problems: Grades 1–2 highlights the use of various strategies and activities and emphasizes the development of proficiency in basic facts and processes for solving word problems. It provides a wide variety of instructional models and explanations for the gradual and thorough development of solving word problems.

The units in this book are designed to match the suggestions of the National Council of Teachers of Mathematics (NCTM). They strongly support the learning of measurement and other processes in the context of problem solving and real-world applications. Use every opportunity to have students apply these new skills in classroom situations and at home. This will reinforce the value of the skill as well as the process.

How to Solve Word Problems: Grades 1–2 matches a number of NCTM standards including the following main topics and specific features:

Problem Solving
Problem-solving approaches are used to investigate and understand mathematical content. Problems are formulated from everyday mathematical situations. The information is designed to develop confidence in using mathematics meaningfully.

Communication
Physical materials, pictures, and diagrams in the text relate to mathematical ideas. The problems reflect upon and clarify thinking about mathematical ideas and situations. This book also relates everyday language to mathematical language and symbols.

Reasoning
The problems include the use of models, known facts, properties, and relationships to explain thinking. The problems also require students to justify answers and solution processes. Patterns and relationships are used to analyze mathematical situations.

Estimation
Some problems explore estimation strategies and help develop the recognition of when an estimation is appropriate.

Number Sense and Numeration
The information included in this book helps students to understand our numeration system by relating counting, grouping, and place-value concepts.

Concepts of Whole Number Operations
This book also develops meaning for the operations by modeling and discussing a rich variety of problem situations. In addition, it helps develop operational sense.

Whole Number Computation
A variety of mental computation and estimation techniques are used in this book. Questions include the selection and use of computation techniques appropriate to specific problems and determining whether the results are reasonable.

Geometry and Spatial Sense
Some mathematical problems ask for students to describe, model, draw, and classify shapes. There are also problems that encourage the recognition and appreciation of geometry in the world.

Measurement
Time measurement is used in everyday situations.

A *word problem* tells about related facts and then asks a question. A *fact* is information that is true. A *question* is something that is asked and needs an answer.

Steps

Introduce these steps and examples to the children.

1. **Read** the word problem together a few times to find out the facts and the question. Look for words in the problem that may give clues whether to add or subtract. This will help to find the answer. Some of the words are:

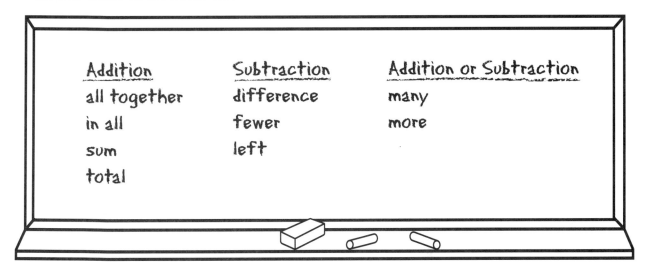

Addition	Subtraction	Addition or Subtraction
all together	difference	many
in all	fewer	more
sum	left	
total		

2. **Name** the words that give you clues whether to add or subtract to solve the problem. Look at the underlined words in the problem below.

> Hanna had 5 strawberries. Her sister gave her 5 <u>more</u>.
> How many strawberries does Hanna have <u>all together</u>?

> Stephen has 9 green grapes. He ate 7 of them.
> How many grapes are <u>left</u>?

3. **Draw** a picture to show the solution, or use real items to perform the problem. In the example below, you could draw apples or use real ones.

> Demarco has 8 apples. He gave 4 to his brother.
> How many apples are <u>left</u>?

Directions: Choose two children to read each of the scripts below. One child is Reader 1 and the other child is Reader 2. Read the script and discuss the solution to each word problem. Look for the words that give clues to add or subtract. Pictures can also be drawn to explain the solution.

Tasty Pieces of Sandwiches!

Reader 1: It's time for lunch! I am so hungry!

Reader 2: So am I! I have a peanut butter and jelly sandwich, and it is cut into 4 pieces so it is easier to eat.

Reader 1: I have a ham and cheese sandwich, and it is cut into 2 pieces.

Reader 2: I wonder how many sandwich pieces we have all together?

Reader 1: If you add 4 + 2, you will find how many pieces there are all together. Do you know the sum?

Reader 2: Yes! We have 6 sandwich pieces, because 4 + 2 = 6. Now let's eat them!

The Disappearing Grapes!

Reader 1: I have a bunch of red grapes, and I will count how many grapes there are: one, two, three, four, five, six, seven, eight, nine, ten, eleven.

Reader 2: I love to eat grapes. Could I please have 4 of them to eat?

Reader 1: Sure, here are 4. Now, I wonder how many are left for me to eat?

Reader 2: You have 7 left, because I subtracted 11 − 4 and got 7.

Reader 1: You are right! I do have 7, and I am going to eat them now!

Reader 2: I ate mine already. I wonder where we can get some more.

Directions: Now, it is time to write your own food addition and subtraction word problems. Remember to write important facts and questions using the words that give clues to add or subtract. Then share your problems with a friend.

Addition (+)

_____

_____

_____

_____

_____

Subtraction (−)

_____

_____

_____

_____

_____

Directions: Everyone loves to eat, but did you know that math can be practiced as you look at your plate? Read each word problem. Then draw a picture on each plate to show how you solved the problem.

Breakfast

1. For breakfast I had a bowl of cereal.
   It had 3 peach slices on it.
   My mom gave me 5 more peach slices.
   How many peach slices in all? _____

Lunch

2. For lunch, I had a sandwich and
   6 carrot sticks on my plate.
   I ate 5 of the carrot sticks.
   How many carrot sticks are left? _____

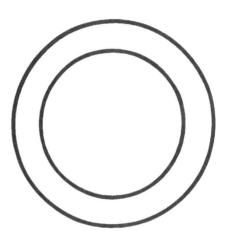

Dinner

My Plate

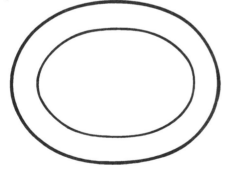

Dad's Plate

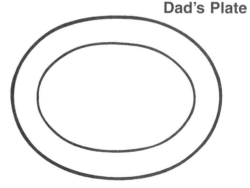

At dinner, my family had a picnic. On my plate, I had 1 hot dog, 1 ear of corn, 12 baked beans, and 5 potato chips. On his plate, my dad had 2 hot dogs, 2 ears of corn, 14 baked beans, and no potato chips.

3. How many more hot dogs did my father have? _____

4. How many more potato chips did I have? _____

5. What was the sum of all the baked beans? _____

6. What was the total of all the ears of corn on both plates? _____

What does communicate mean? To *communicate* means to exchange information and ideas in different ways, such as writing and talking.

Steps

Introduce these steps and examples to the children.

1. **Read** the word problem together a few times. **Write** the facts and questions.

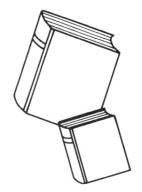

In the month of October, Maria and her brother, Antonio, had to read at least 1 book every day. Maria read 35 books, and Antonio read 31 books.

Who read the most books? _____

How many did he/she read? _____

How many more books did Maria read?

_____

2. **Read** the word problem together again. **Talk** about the words that help to choose addition or subtraction to solve the problem.

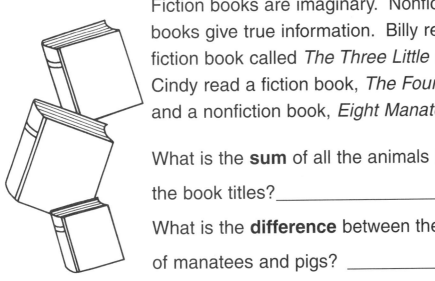

Fiction books are imaginary. Nonfiction books give true information. Billy read a fiction book called *The Three Little Pigs*. Cindy read a fiction book, *The Four Red Hens*, and a nonfiction book, *Eight Manatees*.

What is the **sum** of all the animals in

the book titles?_____

What is the **difference** between the number

of manatees and pigs? _____

How many **more** manatees are there than hens?

_____

3. **Discuss** how math is used in each word problem and how you came to choose addition or subtraction.

Directions: Many people like to read fairy tales. Most fairy tales have math ideas in their stories. Read and talk about the ideas in the titles of these five different fairy tales below. Then read and solve the word problems.

1. Cory read *Nanny Goat and the Seven Little Kids.* Then he read *The Three Billy Goats Gruff.* How many more "kids" were there than "billy goats"? _____

2. Natasha wanted to read *Goldilocks and the Three Bears, The Three Little Pigs,* and *Nanny Goat and the Seven Little Kids* to her cousin. What is the total of all the animals in the titles of these three books? _____

3. Kelly read all five book titles. She found that four books had the sum of 15 animals. Write the titles of these four books.

_____   _____

_____   _____

4. Josh drew 6 swans. Then he drew 7 kids. What is the difference between 7 and 6? _____

Directions: A *pictograph* is a diagram, using pictures, that shows the relationship between a number of objects or people. Study and discuss the pictograph. Then use it to answer the questions.

Cory, Josh, Kelly, and Natasha were supposed to read one book each day for a week. The *pictograph* below shows how many they each read. The *key* shows that each picture of a book represents one book.

A Book a Day

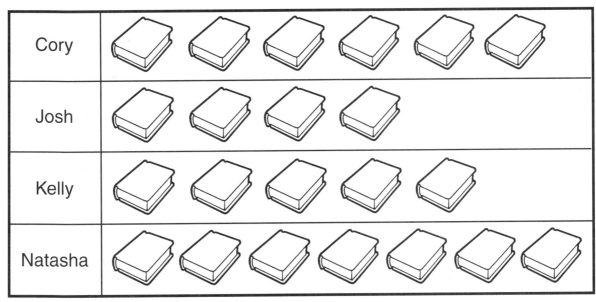

1. Who read the most books?_____ How many did he/she read?_____

2. Who read the least books? _____ How many did he/she read?_____

3. What was the difference between the most books and the least books? _____

4. How many more books did Cory read than Kelly? _____

5. How many books did Cory and Natasha read all together?_____

6. Write each child's name and number of books read, from the greatest to the least.

    a. _____   c. _____

    b. _____   d. _____

Directions: Create a title that includes a number word for each book, for example *My Two Teeth.* Then write an addition and a subtraction word problem using the number words in the book titles. Remember to use the words that give clues to add or subtract.

Addition (+)

_____

_____

_____

_____

Subtraction (−)

_____

_____

_____

_____

What is reasoning? To *reason* is to be able to make conclusions from facts in a logical way. Use the activities below to review facts about geometric shapes, place value, and even/odd numbers.

Geometric Shapes

- Draw and cut out paper shapes of rectangles, squares, trapezoids, and triangles.

- Discuss their different properties. For example, talk about the number of equal and unequal sides.

- Ask questions about the shapes. For example, "How are they alike or different?" or "Which do you see more/less of?"

- Create different patterns using the shapes, such as the pattern below.

- Look around your house, apartment, or neighborhood for examples of shapes.

Place Value

- Review the following fact: the place of a digit in a numeral tells its value.

325

3 has the place value of **hundreds**

2 has the place value of **tens**

5 has the place value of **ones**

- Look for numbers in everyday experiences. For example, point out the values of page numbers in a book or on a restaurant menu.

- Ask questions about the place value of digits, such as "What place do you think the 3 is in the number 36?"

Even/Odd Numbers

- Review the following facts:

> **Even numbers end with 0, 2, 4, 6, or 8.**
> **Odd numbers end with 1, 3, 5, 7, or 9.**

- Help the child to recognize even/odd numbers in everyday life.

- Ask how the child applied reasoning to his or her recognition.

A *neighborhood* is the small area where you live. The part of the neighborhood below is where Manuel, Mark, and Maria live. They like to walk down the street and talk about the different geometric shapes and patterns they see on the houses and buildings.

Directions: Read and solve the word problems.

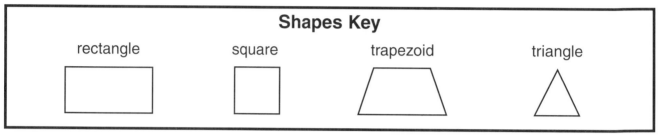

Shapes Key

rectangle   square   trapezoid   triangle

The Neighborhood

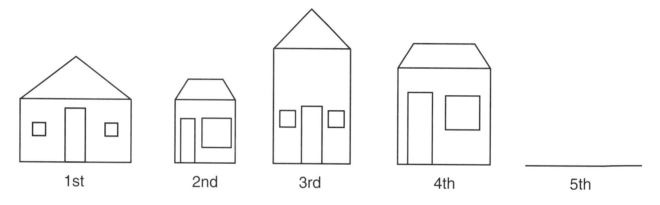

1st   2nd   3rd   4th   5th

1. Maria saw a pattern with the shapes of the roofs. Look at the roofs and draw the pattern and tell why you think this is a pattern.

   _____

2. Manuel looked at the shapes of the buildings under the roofs and saw another pattern. Draw the pattern and tell why you think this is a pattern.

   _____

3. Mark saw a new pattern with the doors. Write about this pattern using two of these words: left, right, or center. Tell why you chose these two words.

   _____

   _____

4. Mark, Maria, and Manuel looked at the windows by the doors and talked about another pattern. Tell a partner why you think this is a pattern.

5. Now, study all the shapes and patterns and draw how you think the 5th building will look. Why do you think the building will look this way? _____

Place value in any number shows the meaning of each digit. The digit's place in the number tells how many it stands for. You can use place-value blocks to show numbers. Look at the example below. These blocks show 126.

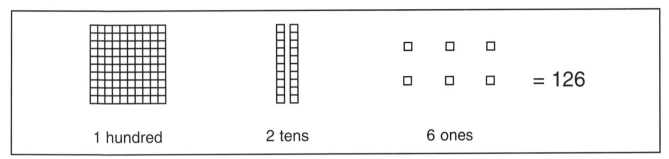

1 hundred   2 tens   6 ones   = 126

Directions: Find the numbers for the place-value blocks below.

1.

□   □   □

□   □   □

□   □   □

_____

2.

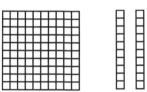

_____

3.

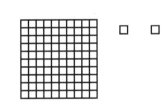

_____

4.

_____

5.

_____

6.

_____

7.

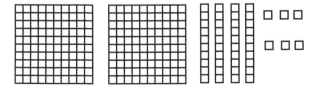

_____

8.

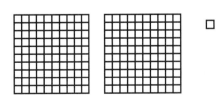

_____

Mark, Manuel, and Maria like to play games with the house numbers in their neighborhood. They write riddles about them. A riddle is a tricky problem with a clever answer.

Directions: Read and study the house numbers and their patterns on First and Second Streets. Then read the riddles, study the clues, and write the answer.

1. Which house am I?
   I am an odd number.
   I have a 7 in the ones place.
   I am less than 99
   and 2 more than 95.
   What is my house number?

   _____

2. Which house am I?
   I am an even number.
   I am more than 663.
   All my digits are even numbers
   and the same digit.
   What is my house number?

   _____

3. Which house am I?
   I am an odd number.
   I have a 6 in the tens place
   and hundreds place.
   The digit in the ones place
   is 3 less than 6.
   What is my house number?

   _____

4. Which house am I?
   I am an even number.
   The sum of my two digits is 17.
   I am less than 100.
   What is my house number?

   _____

5. Follow the patterns and write the two missing
   house numbers on First and Second Streets
   on the empty houses at the top.

First Street

99 97 95 93

Second Street

669 668 667 666 665 664 663

What are math connections? *Math connections* are links that show how math ideas are related instead of isolated. Use the activities below about the odometer and linear measurement to help show how math ideas are related, not isolated.

Odometer

An odometer measures the distance traveled.

- Look at the odometer on a car, van, truck, motorcycle, etc.

- Review the place values of ones, tens, and hundreds on the odometer.

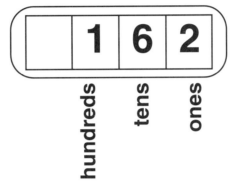

- Notice how the numbers change as the vehicle moves.

- Write down the numbers as they change.

- Review counting by fives from 0 to 100 and the numbers' place values.

Linear Measurement

Linear measurement is measuring the length of a line.

- Discuss linear measurement.

- Practice drawing and measuring lines 1 and 2 inches (2.54 cm and 5 cm) long.

- Draw and connect 1 and 2-inch lines to form a shape like the examples below.

- Look at different kinds of maps and discuss their uses.

- Discuss the use of a map scale and how it helps to measure distances.

Example: •————————• = 10 miles
(1 inch)

- Use the scale to draw lines that equal 20 and 30 miles.

- Practice counting by tens from 0 to 100.

Joshua, Joe, and Jake are triplets. (Triplets are three children of the same birth.) Their parents are driving them to see their twin cousins, Halley and Hanna. (Twins are two children of the same birth.) Halley and Hanna live on a farm 100 miles away. The triplets played some odometer games in the car.

When they left, the odometer read:

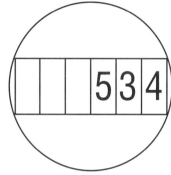

When they arrived, the odometer read:

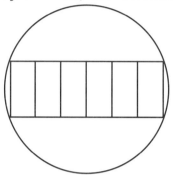

Directions: Read and solve the word problems.

1. The triplets wanted to make a connection between the miles traveled and counting by fives to 100. They made their own odometer for counting. Count by fives and fill in the missing numbers. The first one has been done for you.

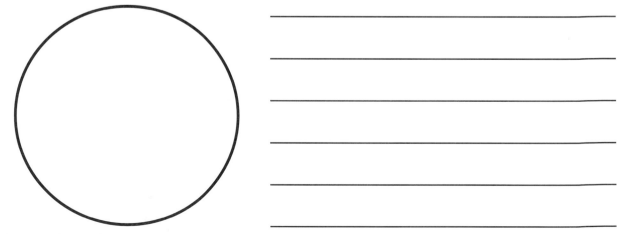

0, 5, (1 0), 15, (☐☐), 25, (☐☐), 35, (☐☐), 45, (☐☐),

55, (☐☐), 65, (☐☐), 75, (☐☐), 85, (☐☐), 95, (☐☐),

2. The odometer read 534 when they left. What number is in the ones place? _____ the tens place? _____ the hundreds place? _____

3. At the top of the page, write the numbers that would be on the odometer when they arrived at their cousins' farm.

4. Draw an odometer that is in your family's vehicle along with its digits. Write a word problem using this numeral. Share it with a partner.

_____

_____

_____

_____

_____

_____

Directions: Look at the map about the triplets' trip from their home to their cousins' farm. Then solve the word problems using the map and the map scale.

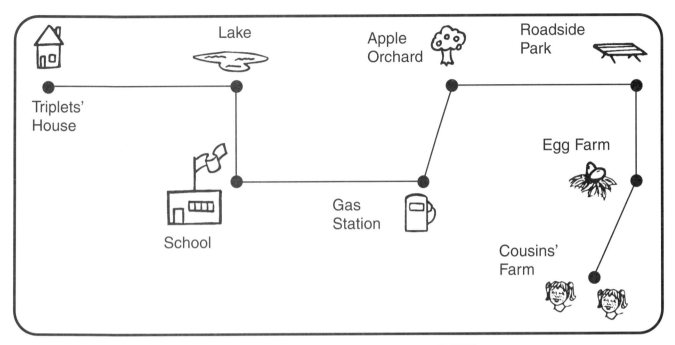

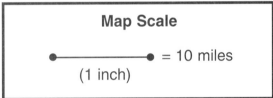

Map Scale

•————————• = 10 miles
(1 inch)

1. How far was it from the triplets' house to the lake? _____ miles

2. How far was it from the lake to the school? _____ miles

3. How many miles was it from the school to the gas station? _____ miles

4. After they stopped at the gas station, how much farther was it to the roadside park where they stopped for lunch? _____ miles

5. They picked apples at the apple orchard and bought eggs at the egg farm. What was the number of miles between the two places? _____ miles

6. What was the total number of miles they traveled? _____ miles. Write how you got this answer.

_____

_____

_____

Directions: The next time you go somewhere in a vehicle, write down the number on the vehicle's odometer on "Odometer One" below before you leave. When you return, write that number on the same odometer on "Odometer Two."

<table>
<tr><td align="center">**Odometer One**</td><td align="center">**Odometer Two**</td></tr>
<tr><td align="center">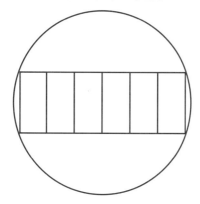</td><td align="center">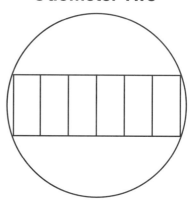</td></tr>
</table>

Directions: Write a word problem using the numbers that you wrote on the two odometers.

_____

_____

_____

_____

_____

Directions: Now, draw a simple map to show how far you traveled using a scale of:

Map Scale

●——————● = _____ mile(s)
(1 inch)

What does estimate mean? In math, *estimate* means to make a close and good guess about the number, size, or cost of something. Use the activities below to help develop your estimating skills.

Size, Number, and Cost

• Go to a food store. Examine and compare the sizes and prices of different, packaged candy.

• Buy two different-sized packages of the same candy. Estimate how many pieces are in the smallest package. Then open and count the pieces. Next, estimate how many pieces are in the larger package. Then open and count the pieces. Compare results.

• Compare the prices of the two packages and the number of pieces in each. Discuss which package was a better buy and how estimating skills were used to make this decision.

• Discuss how estimating is often used in figuring the cost of something.

• After shopping and before emptying a bag, estimate the number of items in the bag. Then count each item. Discuss why the estimation was close or not close.

• Follow the same procedure when there is more than one bag. Compare the totals and why some estimates were different than others.

• Discuss how number and size were used in the estimations, such as fewer large items in a bag or more small items in a bag.

Counting

• Brainstorm different ways to get to 100 by counting and adding, using groups of 5, 10, 20, 25, and 50.

• Count by fives and tens to 100. Add 20 five times to equal 100. Add 25 four times to equal 100. Manipulatives may be grouped to show these processes.

• Practice counting nickels, dimes, and quarters to equal $1.00. For example, emphasize that $1.00 is the same as one dime, three nickels, and three quarters.

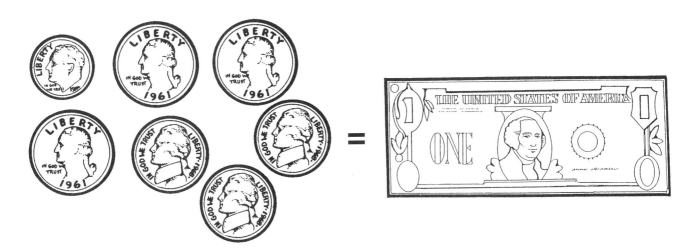

Directions: Read and solve the word problems about the Gummy Squares.

Malika and Matt like to go with their Uncle Dan when he shops at the Dandy Food Market. He always gives them 50¢ to spend on something. Matt and Malika both want to buy some Gummy Squares. They have to decide which to buy—two "pocket boxes" or one "super box."

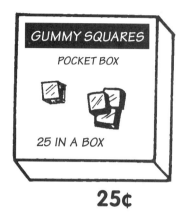

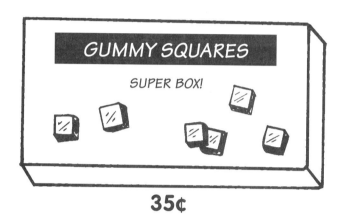

25¢   **35¢**

1. Matt knows there are 25 Gummy Squares in the "pocket box." Estimate how many you think are in the "super box." Tell why you think that. _____

   _____

2. Matt decides to buy one "super box" for 35¢, because he estimates there might be 50 Gummy Squares in the box since it is twice as big as the "pocket box." How much money will he have left? _____

3. Malika spends 50¢ and gets two "pocket boxes." How many Gummy Squares will she get? _____

4. When they get home, Matt counts the Gummy Squares in his "super box." He has 62. Malika has 50. How many more did Matt have? _____

5. Who do you think got more for their money, Malika or Matt? _____
   How did estimating help them to solve their problem? Write about your thinking.

   _____

   _____

   _____

   _____

Directions: Read, estimate, and solve the word problems.

Matt and Malika looked at their favorite fruits at the grocery store. They each have $1.00 to spend on fruit. The signs read:

bananas 10¢ each	oranges 25¢ each	grapes 30¢ a bunch	apples 20¢ each	strawberries 5¢ each

1. Matt loves bananas. Estimate how many bananas he could buy for $1.00. _____

   Tell why you think that. _____

   _____

2. Malika likes apples. Estimate how many apples she could buy for $1.00. _____

   Write why you think that. _____

   _____

3. Matt wants to buy his mother some grapes. Estimate how many bunches he could buy

   for $1.00. _____ Tell why you think that. _____

   _____

4. Malika decides to spend 50¢ on some strawberries. Estimate how many she could buy.

   _____ Estimate what fruit she could buy with the other 50¢. List two or three

   choices. _____

   _____

5. If you had $1.00, write what fruit you would buy. _____

   _____

   _____

Directions: Go shopping at the supermarket with an adult. Estimate how many different things will be put in your basket or shopping cart. Then count the different things you put into the basket or cart. Before you check out with the cashier, do the following activities and record the information below:

- Estimate how many bags will be needed to hold all the items.
- Estimate how many things will go into the first bag.
- Estimate how many things will go into the second bag.

	My Estimate	Correct Number
1. How many things were in the basket or cart?		
2. How many bags were needed to hold all the items?		
3. How many things were in the first bag?		
4. How many things were in the second bag?		

Now, write a word problem using some of your information above and draw a picture in the bag that shows what happens in your word problem.

_____

_____

_____

_____

_____

What is number sense? *Number sense* means to have a sense of the ways numbers are used every day. Use the activities below to help develop number sense.

Sort, Group, and Count

- Find an assortment of objects that can be sorted and grouped by attributes into sets, such as coins, candy, shells, dry beans, stickers, etc. Sort into groups and count the objects in each group.

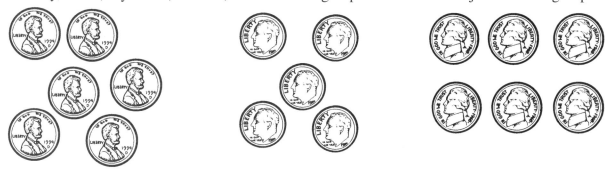

Place Value

- After grouping the objects, regroup each set into tens and ones. For example, group 13 pennies into one set of ten and 3 ones as shown below.

- Say and write two-digit numerals. Tell how many tens and ones in each numeral.

- Find two numerals in everyday life, and tell how many tens and ones are in each numeral. Point out the numbers on signs, house numbers, calendars, TV channels, etc.

Multiple Uses of Numbers

- After sorting, grouping, and counting different objects, brainstorm different meanings for the numbers.

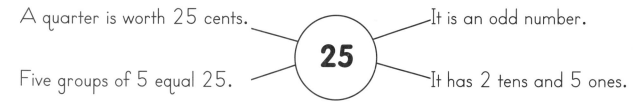

A quarter is worth 25 cents.   It is an odd number.

25

Five groups of 5 equal 25.   It has 2 tens and 5 ones.

- Pick different numbers. Look for them and think how they are used in your daily life. Examples include the following: 25 miles an hour is displayed on a speedometer; a pack of gum is 25 cents; Aunt Ann is 25 years old; or Christmas is December 25.

- Write or tell word problems about the different numbers you find.

6 ▸ Practice • • • • • • • Sorting, Grouping, and Counting

Directions: Read and solve the word problems. Write and draw the solutions to the problems.

B. B. and J. C. collect sports cards. B. B. likes baseball and soccer cards, and J. C. likes football and basketball cards. One day they decided to sort and count their cards.

1. B. B. sorted his cards first. In the first group, he counted 5 baseball cards. In the second group, he counted 7 soccer cards.

   How many cards in all? _____

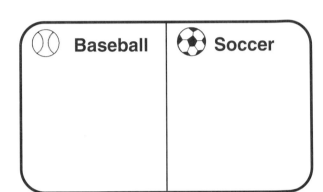

2. J. C. sorted and counted his cards. In the first group, he counted 9 football cards. In the second group, he counted 6 basketball cards.

   How many cards in all? _____

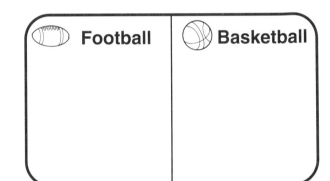

3. In his second group, B. B. counted 9 baseball cards and 8 soccer cards.

   How many cards in all? _____

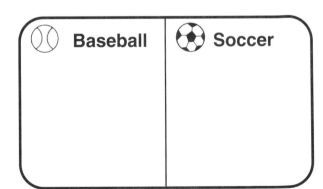

4. When J. C. counted his second group, he had 4 football cards and 10 basketball cards.

   How many cards in all? _____

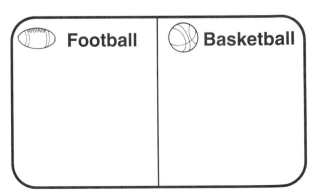

Counting and Grouping to Show Place Value

Directions: Read and study the chart. Then solve the word problems.

The next day B. B. and J. C. decided to sort all of their cards into groups of tens and ones.

B. B.'s Cards			
◯ Baseball		⚽ Soccer	
1	4	1	5
□ □	□ □	□ □ □ □	
□ □	□ □	□ □ □ □	
□ □		□ □ □	
□ □		□ □	
□ □		□ □	

J. C.'s Cards			
🏈 Football		🏀 Basketball	
1	3	1	6
□ □	□ □	□ □ □ □	
□ □ □		□ □ □ □	
□ □		□ □ □ □	
□ □		□ □	
□ □		□ □	

1. B. B. had 14 baseball cards. How many groups of ten? _____ How many ones? _____

2. J. C. had 16 basketball cards. How many groups of ten? _____ How many ones? _____

3. B. B. had 15 soccer cards. How many groups of tens and ones would he have?

   _____ tens _____ ones

4. J. C. had 13 football cards. How many groups of tens and ones would he have?

   _____ tens _____ ones

5. B. B. wanted to know how many cards he had in all. Help him solve this problem. B. B. had _____ cards in all. Tell how you solved this problem.

   _____

   _____

6. J. C. thought he had more cards than B. B. Help J. C. find the sum of all his cards. J. C. had _____ cards in all. Tell how you solved this problem.

   _____

   _____

7. Who had more cards, J. C. or B. B.? _____ Explain your answer.

   _____

   _____

Twenty-nine was a special number for B. B. and J. C. They each had 29 sports cards. Twenty-nine is a numeral that has many math meanings. In the middle of the Number Wheel below is the numeral 29.

Directions: Read and complete each word or math sentence. Use the answers in the list to complete the word sentences.

Answer List								
2	even	9	12	less	21	more	27	odd

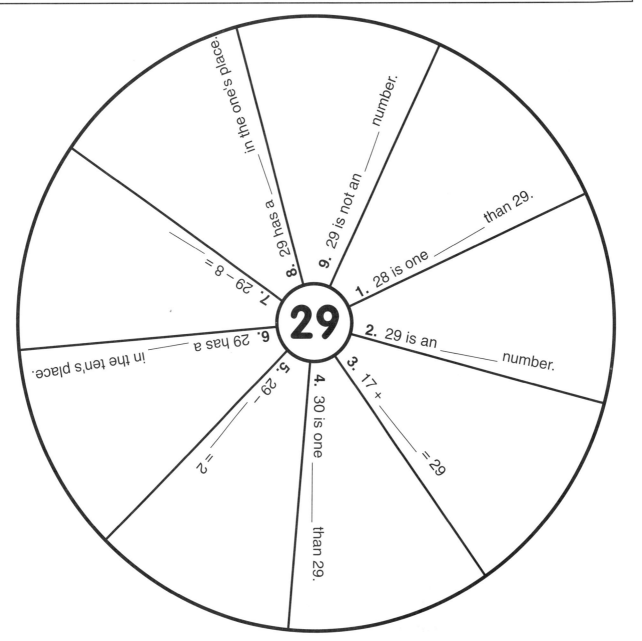

Bonus Question: What is the number of the next page? _____

What are whole number operations? The *whole number operations* of mathematics are addition, subtraction, multiplication, and division. Use the activities below to help develop whole number operations.

Counting

- Count the number of adults, children, and pets in your household and draw a picture to show your findings. How many adults? _____ children? _____ pets? _____

- Tell time on the hour and half-hour, such as 4:00 and 5:30. Understand the difference between times. For example, the difference between 3:00 and 4:00 is one hour.

 difference = 1 hour

- Group pennies by two and count by twos to find the total.

- Count by fives using nickels.

Games

- Play a dice game. Roll one die or a pair of dice a set number of times. Record the numbers rolled and find their sum.

$$\boxed{\vcenter{\hbox{⚂}}} + \boxed{\vcenter{\hbox{⚃}}} + \boxed{\vcenter{\hbox{⚁}}} = 9$$

- Write a variety of one and two-place numerals on index cards. Place the cards facedown in a pile. Take turns drawing a set number of cards, and find their sums.

- Place a set number of objects in a container. Then have other people estimate how many there are.

- Place an amount of pennies on a table, desk, floor, etc. Within 30 seconds, pick up as many as you can. Then count how much money you have.

- Write or tell word problems and number sentences that relate to these or other games.

Graphs

- Find and read bar graphs on the computer, in newspapers, books, or magazines, or create your own.

- Write or tell word problems using the graph's information.

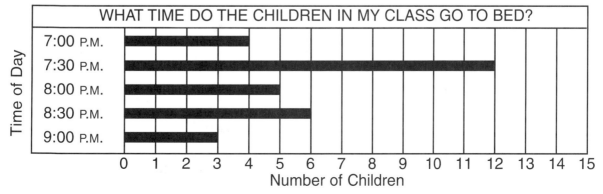

Demetrius and his family are planning his seventh birthday party. He decides to invite 5 girls and 6 boys.

Directions: Help them plan the party as you solve these word problems. Draw a picture to show how you solved each problem. When you are finished, complete the table.

Demetrius' Birthday Party			
Number of Children (including Demetrius)		Cups of Lemonade	
		Cost of Balloons	
Time Length of Party		Number of Prizes	

1. Demetrius needs to know how many children, including himself, will be at the party. How many boys? _____ How many girls? _____

2. The party will start at 1:00 and end at 4:00. How many hours will it last? _____

3. Each child will get two cups of lemonade. How many cups will be served? _____

4. Each child will get a balloon. How many balloons are needed? _____ If they are 5¢ apiece, how much will the total cost of balloons be? _____

5. They will play 10 different games. For each game, there will be 2 prizes. How many prizes will they need? _____

Everyone at Demetrius' birthday party had fun playing the games. The four favorite games were "Dig Those Dice," "Gummy Guessing," "Counting Cards," and "Pickin' Up Pennies."

Directions: Read the directions for each game and solve the word problems.

Dig Those Dice

Each player rolls the die 4 times and adds the four numbers to find a sum. The highest sum wins.

1. Carol rolled

   What was her total score? _____

2. Jaron rolled

   What was his total score? _____

   Write a number sentence to show how you solved this problem. _____

   _____

3. What was the difference between the two scores? _____

Gummy Guessing

A certain number of gummy candy insects are in a jar. Each player estimates how many are in the jar. The player with the closest guess wins. There are 89 insects in the jar. Kelly guessed 81. David guessed 65. Ashley guessed 73. Terrel guessed 80.

1. Write the guesses in order from the highest to the lowest. _____

   _____

2. How many more did Kelly need to get 89? _____

   Write a number sentence to show how you solved this problem.

   _____

Counting Cards

Place the cards with their numbers facedown. Each player takes 4 cards. The player with the highest sum wins.

1. Alecia's cards were: 10 14 21 33

   What was her sum? _____

2. Tom's cards were: 43 11 14 20

   What was his sum? _____

3. The winner was Alex. He had 99 points. What was the difference between Alecia's and Alex's sums? _____

   Write a number sentence to show how you solved this problem.

   _____

Pickin' Up Pennies

Players sit in a circle on the floor. In the middle are many pennies. At the sound of "Go," everyone picks up pennies. When all pennies are picked up, the game is over. The winner has the most pennies.

1. Chinn picked up 57 pennies. How much money did he have? _____

2. Kiah picked up 56 pennies. How much money did she have? _____

3. Demetrius picked up 42 pennies. How much did he have? _____

4. Write two number sentences to tell how many more pennies Chinn had than Kiah and Demetrius.

   _____

   _____

Directions: Complete "The Birthday Graph" by asking 20 or more friends, relatives, etc. the month of their birthdays. Color in one space for each birthday. Then write one addition and one subtraction word problem about the graph.

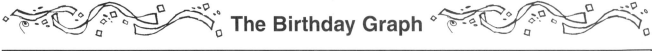 **The Birthday Graph**

January											
February											
March											
April											
May											
June											
July											
August											
September											
October											
November											
December											
	1	2	3	4	5	6	7	8	9	10	11

Addition (+)

_____

_____

_____

Subtraction (–)

_____

_____

_____

What does it mean to compute whole numbers? To *compute whole numbers* means to find an answer through addition, subtraction, multiplication, and division of whole numbers. Use the variety of mental computation techniques below to develop basic fact proficiency. Also, use the activities to help select computation techniques that are appropriate to specific problems and help determine whether the results are reasonable.

Mental Computation Techniques

- Practice adding double numbers using manipulatives.

- Using manipulatives, continue the addition of doubles. Estimate how the sum would change if 1 more was added to one of the addends.

> **If 8 + 8 = 16, then 8 + 9 = 17.**
>
> **If 7 + 7 = 14, then 8 + 7 = 15.**

- Write and draw pictures of doubles and also doubles plus one problems.
- Learn the sum of doubles from 3 to 20 to help in quick mental computation.

Select Computation Techniques

- Review the math terms related to addition and subtraction.

Addition		Subtraction	
4	addend	8	minuend
+ 4	addend	– 4	subtrahend
8	sum	4	difference

- Introduce a variety of real-life word problems using addition and subtraction.

- Explain the process in solving the problems by using manipulatives, drawing pictures, using written or oral expression, etc.

- Read a story and create different addition and subtraction problems related to the story. Discuss whether the results are reasonable.

- Solve a problem mentally. Then solve the problem with a calculator. Compare results. Were they the same or different?

The Dinos wear gray jerseys. A jersey is a shirt with a number to identify the players when they play a game.

Directions: Read the numbers on the jerseys and use them to solve the word problems.

Rocco
32

Ian
48

Jose
73

Willie
15

Skip
87

Chang
20

1. Write an addition problem with the sum of 68. _____

2. Write a subtraction problem with the difference of 14. _____

3. Write a number sentence that tells how much more Skip's number is than Jose's. _____

4. What is the total of Willie's, Chang's, and Rocco's numbers? Write a number sentence that solves this problem. _____

5. Find the total of the largest number and the smallest number. Show this in a number sentence. _____

6. Write a number sentence that tells how much more the largest number is than the smallest number. _____

8 ▶ Practice •••••••••••••••••• Adding the Score

Rocco and Willie like to play football with their friends. They call their team the Dinos. Rocco's younger brother keeps the score for their team. There are four quarters in a football game. One touchdown is 6 points. One extra point can be scored by kicking the ball over the goal post after a touchdown. So, 6 points = a touchdown, and 7 points = a touchdown and one extra point.

Directions: Study the scores for each quarter in the games. Then solve the word problems.

Quarter	1	2	3	4
Game 1	7	6	6	7

Quarter	1	2	3	4
Game 2	12	13	13	12

Quarter	1	2	3	4
Game 3	20	21	20	21

1. In Game 1, the Dinos' score has 2 sets of double numbers: 6 + 6 = _____ and 7 + 7 = _____.

   What is the total score of these doubles? _____ + _____ = _____

2. You know that 6 + 6 = 12 and 7 + 7 = 14. The score can be added in another way, 6 + 6 = 12, so 6 + 7 = _____ (doubles plus one). Show how you would calculate the total score using doubles + one.

   _____ + _____ = _____
   (Doubles plus ones)

3. Write two ways you can calculate the Dinos' final score in Game 2, using doubles and doubles plus one.

   Doubles

   _____ + _____ = _____

   _____ + _____ = _____

   Total _____

   Doubles Plus One

   _____ + _____ = _____

   _____ + _____ = _____

   Total _____

4. Write two ways you can calculate the Dinos' final score in Game 3.

   Doubles

   _____ + _____ = _____

   _____ + _____ = _____

   Total _____

   Doubles Plus One

   _____ + _____ = _____

   _____ + _____ = _____

   Total _____

Directions: Write a math addition or subtraction word problem about each picture.

Jersey Numbers

Final Score	
Dinos	36
Whales	24

_____

_____

_____

_____

_____

_____

_____

_____

_____

_____

_____

_____

The Scoreboard

	1	2	3	4
Dinos	13	14	14	13
Ducks	12	13	12	13

_____

_____

_____

_____

_____

_____

What are plane shapes? *Plane shapes* are geometric shapes that have flat surfaces, such as a square, rectangle, circle, triangle, hexagon, trapezoid, octagon, etc. Use the activities below to help describe, recognize, model, draw, and classify geometric plane shapes.

Classifying Plane Shapes

- Look at pictures of plane geometric shapes. Recognize the characteristics of each shape. Count the number of lines/sides and angles/corners. Look at the sample.

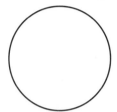

A circle is round.

It has no sides.

It has no corners.

- Discuss how some shapes are alike or different.

- Practice drawing and cutting out these shapes to use as manipulatives.

- Create word problems using the shapes like the example below.

A hexagon has 6 sides, and a triangle has 3 sides.

How many more sides has a hexagon than a triangle?

Create Shape Pictures

- Cut out plane shapes or use pattern blocks or tangrams. Practice fitting the sides together to form complex shapes and designs.

cat standing

duck

- Use paper shapes to design patterns. Example: ▭ △ ○ ▭ △ ○

- Examine quilts, clothing, etc. to see how the different shapes are arranged to form a design.

Geometry in the World

- Look for geometric plane shapes in your home/neighborhood, on the TV/computer, at the store/park, etc.

- Use magazines, newspapers, the Internet, etc. to find pictures with geometric shapes. Cut them out and sort them by shape.

- Use the pictures to make a "shape mosaic," a large picture made up of smaller pictures.

Directions: Suzie and Sylvester Shape and their dog, Spots, have many different shapes. Study their shapes and names. Solve the word problems about Suzie, Sylvester, and Spots. Use the shapes key at the bottom of the page to help you.

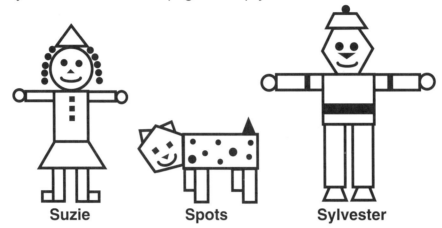

Suzie   **Spots**   **Sylvester**

1. Suzie has _____ rectangles.
   Sylvester has _____ rectangles.
   Spots has _____ rectangles.

   How many rectangles in all? _____

2. Suzie has _____ circles. Spots has _____ circles.

   How many more circles does Suzie have? _____

3. Sylvester has _____ trapezoids.
   Suzie has _____ trapezoids.

   What is the total number of trapezoids they have? _____

4. Spots has 1 shape no one else has. What is this shape? _____

5. Spots has _____ triangles. Sylvester has _____ triangles.

   What is the difference between the two numbers? _____

6. Suzie has _____ squares. Spots has _____ squares. Sylvester has _____ squares.

   What is their sum? _____

7. Which shape has not been counted?
   _____

   How many of this shape in all? _____

Shapes Key

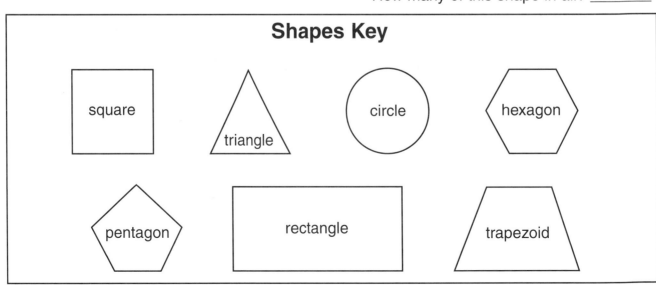

square   triangle   circle   hexagon

pentagon   rectangle   trapezoid

Directions: Rebus stories are fun to read because some pictures are used instead of words. Read "Shapes in Our World." You will find some words are missing. Then reread the story. On the lines, write something in our world that looks like each plane shape. Look at the following example:

I saw a <u>stop sign</u> that was shaped like a ⬡ .

Shapes in Our World

Today I went on a shape hunt with my friends Suzie, Sylvester, and Spots. In the house, I

found a _____ that was shaped like a ☐ and a _____ and a

_____ shaped like a ▭ . Suzie saw a _____ shaped like a △ .

Sylvester looked across the street and saw a _____ shaped like a ⏢ .

At the end of the street, Spots saw a _____ shaped like a ⬡ . We all saw a

_____ , a _____ , and a _____ shaped like a ◯ . It was

hard finding something shaped like a ⬠ , so we pretended we saw a _____ .

We had so much fun that we are going shape hunting again tomorrow!

Directions: Create your own shape picture using the shapes below. Make copies of this on white or colored paper. Cut them out and glue them onto a blank sheet of paper to form a picture, design, etc. Then write or tell word problems about your "Shapes Masterpiece."

Directions: In word problems, it is very important to know basic concepts. Use the following brain teaser to review basic concepts of time. List each word in the Time Box from the shortest time span to the longest. Then write how long each time span is.

TIME BOX			
century	day	hour	minute
month	second	week	year

TIME SPAN	**HOW LONG IS IT?**
1. _____	_____
2. _____	_____
3. _____	_____
4. _____	_____
5. _____	_____
6. _____	_____
7. _____	_____
8. _____	_____

Now, write two word problems using some of this information.

9. _____

_____

_____

10. _____

_____

_____

Directions: Study the sequence of shapes in each pattern. Then draw the next shape that would follow in each series.

1. □ ▱ △ □ ▱ △ □ ▱	
2. ○ □ □ ○ □ □ ○ □	
3. △ ♡ △ ♡ △ ♡ △ ♡	
4. ◇ ◇ ◇ ○ ○ ○ □ □	
5. ♡ △ ♡ ▽ ♡ △ ♡ ▽	
6. ◇ ♡ □ ♡ ◇ ♡ □ ♡	
7. ○ ♡ ♡ △ △ △ ☆ ☆ ☆	
8. ♡ ♡ □ □ ♡ ♡ △ △ ♡	
9. △ △ ☆ △ △ ♡ △ △ ☆ △	
10. □ ▭ □ ▭ □ ▭ □ ▭ □	

Directions: Write the next two numbers in each series. Then write two different number patterns of your own at the bottom.

1. 1, 2, 3, 4, 5, 6, 7, 8, ___ , ___

2. 2, 4, 6, 8, 10, 12, 14, 16, ___ , ___

3. 1, 3, 5, 7, 9, 11, 13, 15, ___ , ___

4. 1, 2, 3, 5, 6, 7, 9, 10, ___ , ___

5. 1, 1, 2, 2, 3, 3, 4, 4, ___ , ___

6. 9, 8, 7, 6, 5, 4, 3, 2, ___ , ___

7. 5, 10, 15, 20, 25, 30, 35, 40, ___ , ___

8. 1, 4, 7, 10, 13, 16, 19, 22, ___ , ___

9. 10, 20, 30, 40, 50, 60, 70, 80, ___ , ___

10. 18, 16, 14, 12, 10, 8, 6, 4, ___ , ___

11. 0, 1, 0, 2, 0, 3, 0, 4, ___ , ___

12. 5, 4, 3, 2, 1, 5, 4, 3, ___ , ___

13. 3, 1, 4, 5, 9, 14, 23, 37, ___ , ___

14. 3, 6, 9, 12, 9, 6, 3, 6, ___ , ___

15. _____

16. _____

Directions: Answer the 12 math trivia questions. Then write two more math trivia questions and share them with a partner.

1. How many minutes are in two hours? _____

2. How many dimes are in a dollar?_____

3. What is a three-sided figure called?_____

4. What do you call the answer in a subtraction problem? _____

5. How many things are in one dozen? _____

6. How many sides does a hexagon have? _____

7. How many months are in half a year? _____

8. How many sides are on a die?_____

9. What do you call the answer in an addition problem? _____

10. How many days are in two weeks? _____

11. Which plane shape has five sides? _____

12. How many hours are in a day? _____

Now, write two more math trivia questions and answer them.

13. _____

_____

14. _____

_____

Materials

- computer with paint software
- printer

Before Using the Computer

- Children should be familiar with paint and draw tools in the software program and know how to use the printing function.
- The teacher or parent should change the animals to be drawn to correspond with the stamps that are available in the software program.

At the Computer

1. Begin with a blank document in the paint software program.

2. Tell the children that they will use the stamp function or paint and draw tools to draw animals.

3. Ask the children to stamp or draw three ladybugs beside each other on the screen.

4. Ask them how many legs a ladybug has. Ask the children if all the ladybugs' legs are visible on the screen. If they are not, have the children draw them in.

5. Ask the children how they might find out how many legs three ladybugs have all together.

6. Have the children count the legs on the three ladybugs and type their answers on the computer.

7. Continue the activity with the following word problems:

    Draw 4 bluebirds. How many wings are there on 4 bluebirds?

    Draw 6 puppies. How many ears are there on 6 puppies?

    Draw 5 frogs. How many legs do 5 frogs have?

    Draw 7 cats. How many tails can wag when 7 cats are in the room?

8. Have the children print their screens of stamped and drawn objects with the answers.

9. After the children have printed the activity, ask them to explain how their answers were reached using the pictures.

Extensions

- Children can create problems for other children to solve.
- Children can create seasonal or theme-related problems to solve.

Materials

- computer with paint program
- printer

Before Using the Computer

- Children should be familiar with the draw and paint tools in the software program and know how to print.

- The teacher or parent should change the candy to be drawn to correspond with the stamps that are available in the software program.

At the Computer

1. Display a blank document in the paint program on the monitor.

2. Tell the children they are going to solve word problems about the candy store using the computer to draw the pictures and write their answers.

3. Read the following word problem to the children: "Paul's mother bought him 3 candy canes at the store. Paul bought 2 grape lollipops too. Later, his uncle gave him 5 jawbreakers. He put all the candy in a bag. How many pieces of candy does Paul have?"

4. Ask the children to draw the candy pieces described in the problem and type their answers. Children may use the stamp feature or any of the paint tools available to draw a simplified picture of the candy pieces. Help them if necessary.

5. Continue reading the following word problems:

    Linda got 5 chocolate drops from her Aunt Lisa. Tommy gave Linda his coconut ball, and Jose gave her 4 lemon cubes. Linda bought 2 peppermint patties from the candy store. How many pieces of candy does Linda have all together?

    Carlita gave Paul 4 lemon drops, Charlie 6 cinnamon drops, and Marcie 7 red hot candies. How many candies did she give away?

    Tabitha decided to split her candy with Maria. She gave Maria 2 malted milk balls, 5 jelly beans, and 4 jawbreakers. If she split her candy evenly, how much candy did she have to start with?

    The candy store owner gave Larry and David a bag of candy. Each boy had 2 chocolate bars, 5 gummy worms, 3 caramels, and 2 licorice sticks. How many pieces of candy did the store owner give away?

6. Have children print their screens of stamped and drawn objects with the answers.

7. After the children have printed the activity, ask them to explain how their answers were reached using the pictures.

Extensions

- Children can create candy word problems for other children to solve.

- Children can create seasonal or theme-related problems to solve.

Page 8
1. 8 peach slices
2. 1 carrot stick
3. 1 hot dog
4. 5 potato chips
5. 26 baked beans
6. 3 ears of corn

Page 9
1. Maria, 35 books, 4 books
2. 15 animals, 5, 4

Page 10
1. 4 more "kids"
2. 14 animals
3. Goldilocks and the Three Bears, The Three Little Pigs, The Six Swans, and The Three Billy Goats Gruff
4. 1

Page 11
1. Natasha, 7 books
2. Josh, 4 books
3. 3 books
4. 1 book
5. 13 books
6. a. Natasha-7, b. Cory-6, c. Kelly-5, d. Josh-4

Page 14
1. triangle, trapezoid, triangle, trapezoid
2. rectangle, square, rectangle, square
3. center, left, center, left
4. one window on each side of door,
   one window on right side of door,
   one window on each side of door,
   one window on right side of door
5. Answers will vary.

Page 15
1. 9
2. 13
3. 35
4. 120
5. 118
6. 102
7. 246
8. 201

Page 16
1. 97
2. 666
3. 663
4. 98
5. 100 and 670

Page 18
Odometer: 634
1. 20, 30, 40, 50, 60, 70, 80, 90, 100

2. 4, 3, 5
3. 634
4. Answers will vary.

Page 19
1. 20 miles
2. 10 miles
3. 20 miles
4. 30 miles
5. 30 miles
6. 100 miles

Page 22
1. Answers will vary.
2. 15 cents
3. 50
4. 12
5. Matt, answers will vary.

Page 23
1. 10 bananas, answers will vary.
2. 5 apples, answers will vary.
3. 3 bunches of grapes, answers will vary.
4. 10 strawberries, answers will vary.
5. Answers will vary.

Page 26
1. 12 cards
2. 15 cards
3. 17 cards
4. 14 cards

Page 27
1. 1 ten, 4 ones
2. 1 ten, 6 ones
3. 1 tens, 5 ones
4. 1 tens, 3 ones
5. 29, answers will vary.
6. 29, answers will vary.
7. Neither. They both had 29.

Page 28
1. less
2. odd
3. 12
4. more
5. 27
6. 2
7. 21
8. 9
9. even
Bonus: 29

Page 30
1. 12 children, 7 boys, 5 girls
2. 3 hours
3. 24 cups
4. 12 balloons, 60 cents
5. 20 prizes

Page 31

Dig Those Dice

1. 16
2. 13, 2 + 6 + 4 + 1 = 13
3. 3

Gummy Guessing

1. Kelly-81, Terrel-80, Ashley-73, David-65
2. 8, 89 − 81 = 8 or 81 + 8 = 89

Counting Cards

1. 78
2. 88
3. 21, 99 − 78 = 21 or 78 + 21 = 99

Pickin' Up Pennies

1. 57 cents or 57¢
2. 56 cents or 56¢
3. 42 cents or 42¢
4. 57 − 56 = 1 or 56 + 1 = 57, 57 − 42 = 15 or
   42 + 15 = 57

Page 34

1. 48 + 20 = 68
2. 87 − 73 = 14
3. 87 − 73 = 14 or 14 + 73 = 87
4. 15 + 20 + 32 = 67
5. 87 + 15 = 102
6. 87 − 15 = 72 or 15 + 72 = 87

Page 35

1. 12, 14, 12 + 14 = 26
2. 13, 13 + 13 = 26
3. 12 + 12 = 24, 12 + 13 = 25,
   13 + 13 = 26, 12 + 13 = 25
   Total 50, Total 50
4. 20 + 20 = 40, 20 + 21 = 41,
   21 + 21 = 42, 20 + 21 = 41
   Total 82, Total 82

Page 38

1. 5, 9, 5, 19
2. 15, 8, 7
3. 3, 1, 4
4. pentagon
5. 4, 1, 3
6. 5, 2, 2, 9
7. hexagon, 1

Page 41

1. second
2. minute = 60 seconds
3. hour = 60 minutes
4. day = 24 hours
5. week = 7 days
6. month = 4 or 5 weeks
7. year = 12 months
8. century = 100 years
9. Answers will vary.
10. Answers will vary.

Page 42

1. triangle
2. square
3. triangle
4. square
5. upright heart
6. diamond
7. star
8. heart
9. triangle
10. rectangle on top

Page 43

1. 9, 10
2. 18, 20
3. 17, 19
4. 11, 13
5. 5, 5
6. 1, 0
7. 45, 50
8. 25, 28
9. 90, 100
10. 2, 0
11. 0, 5
12. 2, 1
13. 60, 97
14. 9, 12
15. Answers will vary.
16. Answers will vary.

Page 44

1. 120 minutes
2. 10 dimes
3. triangle
4. difference
5. 12 things
6. 6 sides
7. 6 months
8. 6 sides
9. sum
10. 14 days
11. pentagon
12. 24 hours
13. Answers will vary.
14. Answers will vary.

Page 45

6 + 6 + 6 = 18 legs
2 + 2 + 2 + 2 = 8 wings
2 + 2 + 2 + 2 + 2 + 2 = 12 ears
4 + 4 + 4 + 4 + 4 = 20 legs
1 + 1 + 1 + 1 + 1 + 1 + 1 = 7 tails

Page 46

3 + 2 + 5 = 10 pieces of candy
5 + 1 + 4 + 2 = 12 pieces of candy
4 + 6 + 7 = 17 candies
2 + 2 + 5 + 5 + 4 + 4 = 22 candies
2 + 2 + 5 + 5 + 3 + 3 + 2 + 2 = 24 pieces of candy